The Saltwater Marsh, A Magical Place

Written by Consie Berghausen
Illustrated by Nina Berghausen

Riverhaven Books

Published in the United States by
Riverhaven Books, Massachusetts.

ISBN 978-1-937588-49-6

Printed in the United States of America
by Country Press, Lakeville, Massachusetts

Designed by Stephanie Lynn Blackman
Whitman, MA

Dedicated to my family,
with love and appreciation

The saltwater marsh,
a magical place,
Both fresh and salt water
meet in this space.
High tide rolls in,
the marsh fills like a lake;
Low tide goes out
and the water escapes.

Out to the sea
with the waves on the beach,
Past fish, crabs, and seaweed
only to reach
Back toward the marsh
with the late changing tide,
Weaving through grass
on its long river ride.

Tides move in and out
every six hours.
Low to high water,
pulled by Moon's power.
Magic because
Within water and grass,
Wonderful creatures
Live, dine, and pass.
At sunrise each day
the marsh comes awake
With rabbits, blue heron,
crows, deer, and snakes,
With ducks, fish, and fiddler crabs
by the shore,
With seagulls and foxes
and so much more.

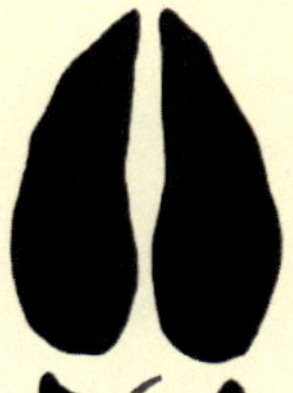

High up in the oak tree, shiny and black,
A scout crow watches and waits for a snack.
Through many bushes a fox slowly creeps,
While back at home all his kits are asleep.

The crow spots fox and starts to caw loudly,
Crow friends join in and start cawing proudly.
They caw together to scare fox away.
They are not friends, and that's how it will stay.

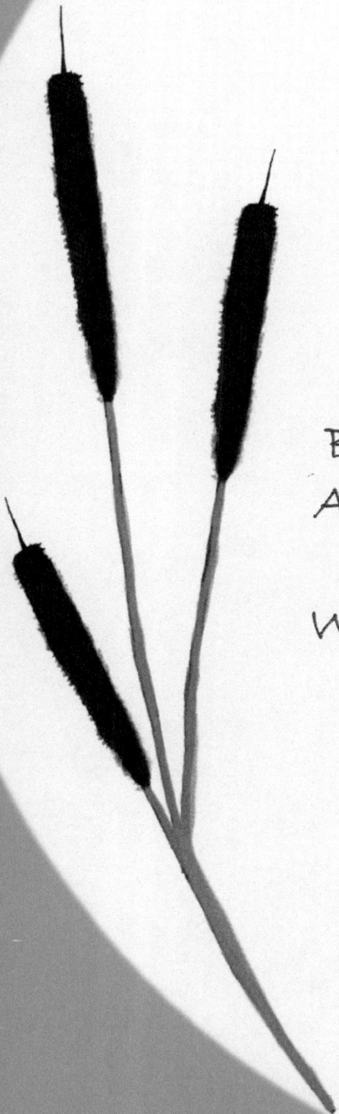

Blue heron flies in and readies to land;
Along water's edge she chooses to stand.
For up to an hour, she does not twitch;
Watching the water, she waits for a fish.

The marsh ducks swim
in a very straight line,
A mother in front
and children behind.
Ducklings have fun
as they paddle and dive,
To stay close to Mother
they quickly strive.

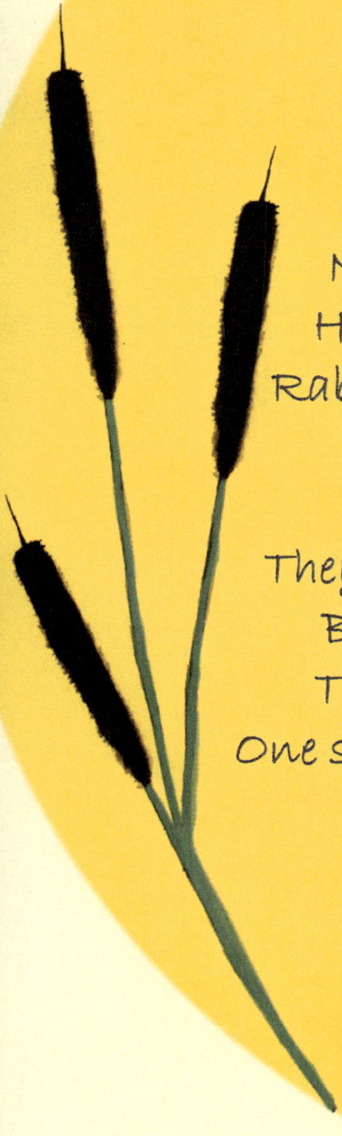

Nearby a deer nibbles grass under trees,
Her newborn fawn close on wobbly knees.
Rabbits watch closely as they hide from view,
They're always alert for predators too.

They don't stay long, away they must hurry,
By water's edge the fiddler crabs scurry.
They are small and each is only an inch,
One small and one large claw, they like to pinch.

On high ocean breezes,
the seagull soars.
From ocean to pond,
on wing they explore.
They fly with the osprey,
herons, and hawks,
Flapping and gliding,
they sail and they squawk.

The marsh day closes,
the sun starts to set.
Home head the animals,
tired and wet.
The red-winged blackbirds
tuck into their beds,
In cattails that grow
near the marsh's edge.

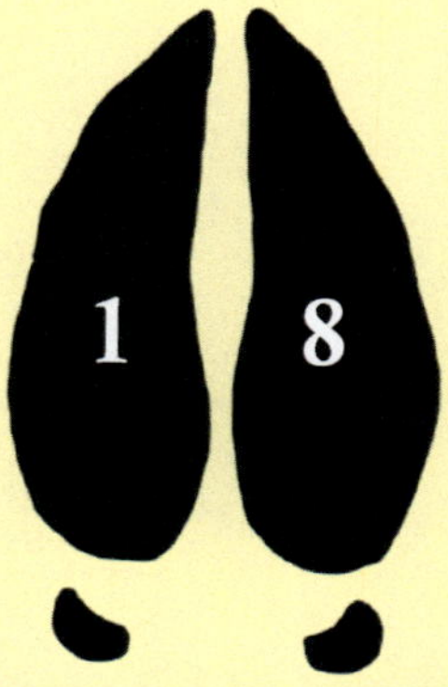

The fox in the den,
the birds in their nest,
Deer, duck, and rabbit
have some place to rest.
Down for the night
to get much needed sleep,
While nocturnal animals
start to creep.

In the marsh, in the dark,
a new show starts:
Raccoons, night herons,
and owls embark.
Safe in the night
they can wander around,
Some thumping, some chirping,
some without sound.

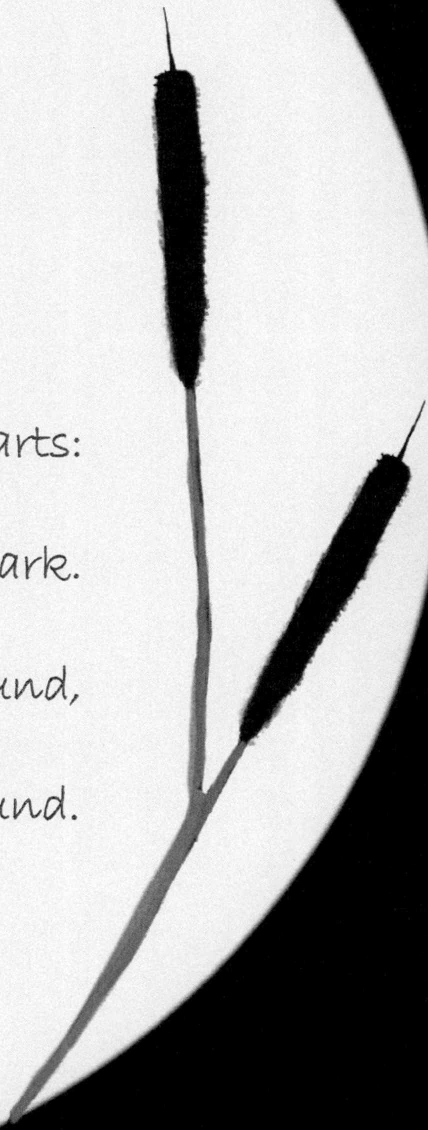

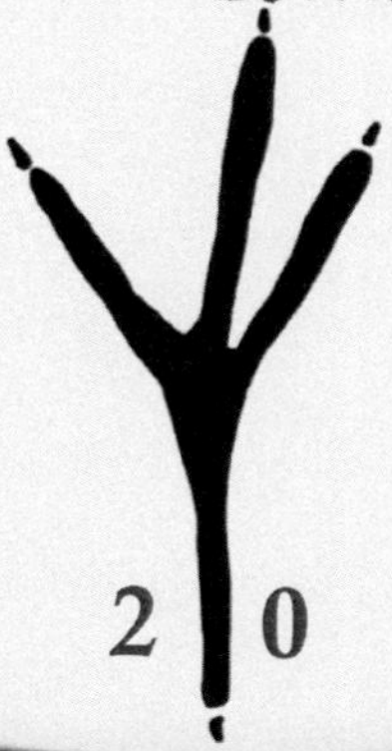

A squeaky brown bat
glides in wondrous loops,
Hunting for bugs
he must dive, spin, and swoop
The mocking bird sings
in many voices;
Stealing bird songs
she has many choices.

A furry raccoon
climbs
down
from
his
tree.
His reflective eyes
allow him to see.
He wanders around
for he has a wish,
To hunt for berries
or maybe a fish.

The barred owl nests
in a tree close to home,
Watching from high
as below creatures roam.
His song in the night
is deep, clear, and true;
He calls in owl language,
"Who cooks for you?"

The coyotes like
to roam in a pack.
They hunt and they play
and they arch their backs.
They howl out loudly
for all friends to hear;
Their yips are quite loud,
So they must be near.

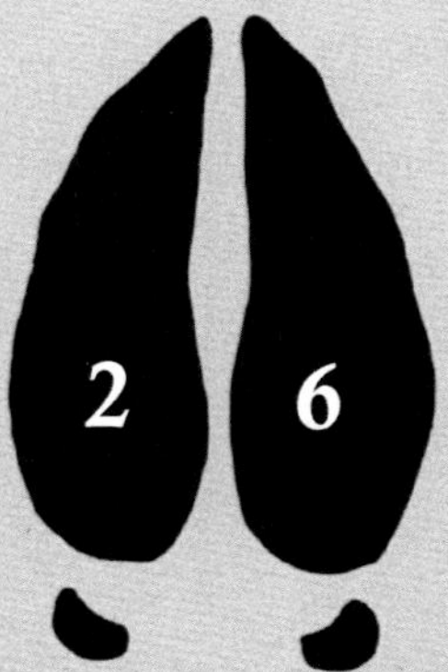

The sky starts to lighten,
stars fade away,
Signaling nature
to start a new day.
Night critters go
and get ready for sleep,
While early birds wake
and begin to peep.

The saltwater marsh,
a magical place,
Both fresh and salt water
meet in this space.
High tide rolls in,
the marsh fills like a lake;
Low tide goes out
and the water escapes.

Out to the sea
with the waves on the beach,
Past fish, crabs, and seaweed
only to reach
Back toward the marsh
with the late changing tide,
Weaving through grass
on its long river ride.

You've reached the tail end
Of our story today,
But, friends, do not fear,
There are more on the way.

Please visit
www.ConsieBerghausen.com
to learn more about her stories,
to download worksheets,
and to see what both author and illustrator
are working on next.

Made in United States
Orlando, FL
12 September 2026

87594397R00024